The Course of Nature
A BOOK OF DRAWINGS ON NATURAL SELECTION AND ITS CONSEQUENCES

That many and grave objections may be advanced against the theory of descent with modification through natural selection, I do not deny. I have endeavoured to give to them their full force. Nothing at first can appear more difficult to believe than that the more complex organs and instincts should have been perfected not by means superior to, though analogous with, human reason, but by the accumulation of innumerable slight variations, each good for the individual possessor.

CHARLES DARWIN

Any view of the Universe that is not strange is false.

NEIL GAIMAN
From "Sandman" #39 © DC Comics. Available in Sandman: Fables and Reflections

The
Course
of Nature

A BOOK OF DRAWINGS
ON NATURAL SELECTION
AND ITS CONSEQUENCES

Amy Pollack

Commentary by Robert Pollack

ISBN 9 781499 122244

Design by Summer J. Hart

To the course of marriage:
Science and Art,
Charles and Emma Darwin,
and our own marriage of fifty years.

TABLE OF CONTENTS

Few people would deny the significance of Charles Darwin's work or the impact that his theories have had on our understanding of history, nature, and our place in the universe. In fact, one would be hard pressed to find a theory that has informed our collective consciousness as much as the theory of evolution by natural selection. Which is why when it came time for the College of Arts and Letters (CAL) at Stevens Institute of Technology to compile a list of works to excerpt for inclusion in a Reader for our new freshman Humanities Colloquium, Darwin's On the Origin of Species was at the top of the list. The expressed goal of the Colloquium was to introduce students to all the humanistic disciplines in order not only to provide them with some familiarity with selected seminal works, but also to instill in them a deep appreciation for the value these works hold when it comes to navigating today's pressing social and moral issues. The faculty involved in the project unanimously agreed that there was no better way to give students a keen insight into the discipline of history and the effects that compelling scientific evidence has on how we understand basic concepts and ideas than a study of the work that forever changed the intellectual landscape of human knowledge. However, inasmuch as the aim of our new course was to introduce students to the humanities, our primary goal in having them read Darwin was to teach them about history, or perhaps the sociology of ideas, and not necessarily biology.

Nevertheless, as the course progressed, we soon realized that a deeper understanding of the biology was critical in order to appreciate fully Darwin's revolutionary work; and that a keener insight into things like adaptation, natural selection, genetic coding and mutation, along with an awareness of how the various natural processes worked, were all sine qua non for a proper comprehension the theory of evolution and the lasting impact it has had on our understanding of both nature and our place in it. But since we were all humanists, we could only hope our students would seek out the more scientific knowledge on their own. Which is why we were so delighted when our dear friend and CAL advisor Bob Pollack suggested that his wife Amy's drawings depicting "evolution and its consequences" be used as a companion text for our course. Here before us was an opportunity not only to have the biology explained by one of the most renowned experts in the field, but to have it done in a way that would make use of a mode of presentation that is most relevant to CAL – that is, the work of art. It seemed that our partnership with the Pollacks on this project was a perfect – or as Bob likes to say, a "pre-adapted" – match. Bringing together the drawings of artist/scientist Amy Pollack with the narrative of scientist/artist Bob Pollack has produced a work at the critical intersection of science, technology, the humanities, and arts. In that respect the Pollacks have provided us with a true gift – of their talents, their wisdom, their insights, their generosity, and their inspiration.

The quantum physicist Niels Bohr once noted that both science and art are concerned with creating images in order to explain and express reality. That is precisely what Amy Pollack's drawings do. They express ideas and concepts that oftentimes cannot be fully captured by mere words; in that respect they provide the perfect complement to the scientific theory. And as Bohr might further note, as complementary, both science and art are requisite for a full explanation of natural phenomena. It is worth noting that Bohr further remarked that in the end, however arcane or even counterintuitive our scientific ideas might be, we must be able to explain them in terms of everyday language. The imagery that Amy Pollack uses in these drawings enables her to do exactly that. By taking notions and ideas from history, literature, art and architecture, culture and folklore, as well as science itself, Amy has created works that summon the viewer to enter the world of scientific theory in a way where it becomes a familiar place. The drawings are intended not merely as accompaniments to the text, but rather as an enhancement of what is being explained. In turn, the role of the text is "to elaborate on what each drawing instantaneously conveys." In each case readers should begin with the drawing on the left, immersing themselves in the richness and power of the ideas encoded within. Doing so will allow Amy's drawings to achieve their ultimate goal, which is to take the science and "humanize it up" rather than "simplify it down."

I believe the most compelling aspect of this moving and enlightening work is found in its ability to refute that common misconception that Darwin's theory has left human life bereft of meaning, and that if correct, natural selection eliminates the uniqueness of human existence. One finds a welcome riposte throughout this work, but above all in the chapter titled "Being Human." After having noted that it is indeed the case that "as a species, we're one of tens of millions of species alive today, an ordinary living twig on the great bushy top of the long spindly tree of life," Bob goes on further to recognize our unique ability to use those "human-specific mutational differences to construct mental worlds of novel complexity"--something only we can do. Thus it is humans alone who are able to "speak thousands of languages, invent music, art, science, law, philosophy, economics, sports and games, and religions." The key, of course, is to do all this in ways that allows us to benefit from our inherited capacity for love--all forms of love--that is encoded in us at birth, as if a "baseline gift of nature."

Amy and Bob Pollack capture this important insight regarding the uniqueness of human existence most poignantly toward the end where we find a drawing of a turntable playing "the recording of life (that) has played out through the amplifier of natural selection for more than four billion years." As Bob goes on to note, "our difference is, of course, that our mental capacities and our capacities for joint action do give us a unique extra capacity to control our future, a capacity we do not share with any other known creature on Earth. We are not merely a song on this record; we can make our own recordings. Let's hope we use that power wisely." I am confident that reading this work will inspire readers to heed this wish and compose their lives wisely, prudently, and harmoniously with others.

Lisa M. Dolling

What can we know about the whole Universe? What is life? How do I fit in? Every thinking person asks these questions, and for most of us the answers remain elusive. That's not because the questions are hard, nor because the answers are vague. It is because the answers we find hardest to toss away – the ones derived from experiment – are unexpected, unintuitive and really hard to accept.

I am a scientist, and Amy is an artist. We've had the good fortune to gain enough patience from each other to persist in confronting with the yearnings of our own hearts these facts that resist disproof. This book began as a project to bring together Amy's drawings and my arguments for a text supplementing the Stevens Institute of Technology CAL Colloquium reading by Charles Darwin.

When my colleague at Columbia University, Psychology Professor David Krantz, called Amy out of the blue to offer her a significant and unsolicited grant to publish her drawings this book needed only to be put together. For that, my colleague at the Columbia University Seminars, Summer Hart, emerged as a designer. The format of the book is very simple. Each pair of facing pages has a drawing by Amy, and an explanation by me. My task has been to elaborate on what each drawing instantaneously conveys.

The pictures are sorted into ten chapters. These are at once the ten rooms of a gallery of drawings for anyone interested in novel representations of difficult concepts, and the ten sessions of a course in science for students of any age who are willing to be taken by surprise at the workings of nature.

The chapters follow a line of argument, though each chapter is self-sufficient for a class or set of classes on their subjects. The argument goes from the largest context for life on Earth (chapters 1 and 2), to the paradoxical properties of the self-replicating, information-containing molecule DNA (chapter 3), to the ways in which DNA expression and DNA regulation construct a living thing from the information in its DNA (chapters 4 and 5), to the particularities and advantages of sex (chapter 6), to the paradox of DNA similarity among all living lifeforms (chapter 7), to the resolution of this second paradox by Charles Darwin (chapter 8), to the particularities of humanity's origins as explained by Darwinian Natural Selection (chapter 9), to the conclusion that our mental worlds give us a unique place in all of nature, marked by our novel freedom to act for better or worse (chapter 10).

Robert Pollack

I – *Where's Here?*

The gift and the burden of science are the same: to present us all with explanations of nature that have survived all attempts at disproof. The most stunning example of how a scientific explanation may survive disproof without in any way becoming less strange, is the conclusion that time and space both began at a single point and at a single time.

The data we have, together with fairly well-established astrophysical theories based on Einstein's general relativity theory, strongly suggest that there was an instant about 13.7 billion years ago at which both time and space began as a singularity, a point containing all of space. So everyone's first questions – what was around that point and what was before it? – are impossible to answer. There was no "around," nor was there a "before." At that point, everything was there, and time had just begun.

How did things get from that instant to here and now? To begin with, as time began that point expanded far more rapidly than the speed of light. As it expanded, the average temperature of the universe cooled down enough to permit the parts of atoms to assemble, and then to be clustered together by an attractive force – gravity – that draws any collection of atoms to each other in proportion to their masses.

Even as gravity pulled together the first stars and then clumped them into the first galaxies, the universe itself, what Einstein called the "space-time continuum," continued to expand and is expanding even now. We are constantly getting further away from all other parts of the Universe not because we are traveling away, but because with its expansion, the distances between any two parts of the Universe become bigger. We are in the three-dimensional version of what dots on the surface of a balloon experience as the balloon is being blown up.

II – *Earth's Fruitful Peel*

Life covers the planet Earth, but barely. We can find living things a few miles deep into the oceans, and a few miles up into the atmosphere, and on the surface of all land. But our planet has a diameter that measures in the thousands of miles. The part of Earth that holds life, called the biosphere, is restricted to a very thin wrapping that lies over, on and in the Earth. So far as we know, that wrapping is the entirety of life's presence in the Universe. We may think of ourselves as dominating the planet, but the reality is that especially to the extent that we dominate it, we must somehow continue to fit into this thin membrane on a planet, a membrane fragilely balanced to permit our existence.

How thin is it? When dealing with ideas that require expression of very large differences in size or any other measure, it is useful to have a language of approximation. We can easily round off a measurement in any units to the nearest factor of ten. Since our number system adds a digit to the size of a number every time it increases by ten-fold, we can reduce the measurement to the number of zeroes in it. This is called the exponent: 10^6 is shorthand for 1 with six zeroes after it, or a million. By convention, 10^0 is 1. So, 10^9 is 1,000,000,000, that is, a billion. That's a thousand times bigger than a million, which is 10^6 or 1,000,000; a million times bigger than 10^3 which is 1,000 or a thousand; and a billion times bigger than 10^0 or 1. Quantities smaller than 1 have negative exponents.

So, in terms of exponents, the biosphere's thickness is only about 1/1,000, or 10^{-3}, times the diameter of the Earth.

Rounded to the nearest factor of ten, and using scientific units, a typical apple is on the order of 10 centimeters in diameter, and its peel is no more than a tenth of a centimeter, or 1 millimeter, thick. That means that the ratio of the peel to the apple is about 10/0.1, or 10^{-2}. That is ten times thicker relative to the apple, than the thickness of the biosphere is, relative to Earth.

Only six of the elements found in and on the surface of the Earth make up almost all of the material of the living world: hydrogen, oxygen, carbon, sulfur, phosphorus and nitrogen. Why is life so prolific within the constraints of this thin membrane, and so absent elsewhere? Because life is made of atoms, and since atoms are so small, life can be made of complicated sets of atoms assembled into very improbable structures. Life on earth turns out to have needed only these atoms and the temperatures found in this thin surface layer.

III – *Deep Water, Small Molecules*

How small are atoms? We know very little about Titus Lucretius Carus: he was born in Rome about 100 BCE, died before he was fifty, and left us only one long poetic essay, *On the Nature of Things*. In that poem Lucretius tries to ease the fear of death by imagining a universe composed entirely of invisibly tiny objects. He calls these atoms, to indicate that they are also irreducible [a-tom/no-parts].

About a hundred years ago – only two thousand years later – scientists testing their ideas about the structure of matter became pretty confident Lucretius had got it at least partly right. All parts of the Universe we can detect – the living and the rest – are made of atoms and the forces and particles that hold atoms together. Real atoms are divisible into smaller parts though; atomic energy and atomic weapons both use that ironic fact.

There are more than a hundred different kinds of atoms, each made of a tiny nucleus holding almost all of the atom's mass, surrounded by a relatively large, spherical cloud of electrons. How small are atoms relative to anything we can see with our own eyes? The smallest unit of distance on a common ruler is the millimeter. The diameter of an atom is approximately 10^7 millimeter; this means atoms are small enough for ten million of them to be lined up in a row inside that smallest unit.

Here's another way to look at how small atoms are. A molecule of water is made of two hydrogen atoms and an oxygen atom, held together by the sharing of their electron clouds: H_2O. If you take a large mug, one that holds about ten ounces, and fill it with water, it will have about 10^{25} molecules of water in it.

It would require about 5×10^{21} mugs of water, each with 10^{25} molecules of water, to fill all the oceans on Earth. So, there are thousands of times more molecules of water in a single mug, than there are mugs of water in all the oceans of the world. Rounding to the nearest exponent, 10^{25} molecules per mug divided by about 10^{22} mugs in all oceans equals about 10^3 times more molecules in one mug than mugs in all the oceans.

So if you were to fill that mug from any ocean, pour it back, mix thoroughly with all the oceans, and then dip and fill the mug again, you'd be certain of getting a few thousand of the water molecules from the first mug, in the mug the second time. That's how small the atoms are that make a single molecule of water, or of any other part of the universe.

EACH CELL HAS 10^{15} ATOMS

$10 \times 10 \times 10 \times 10 \times 10 \times 10 \times 10 \times 10 \times 10 \times 10 \times 10 \times 10 \times 10 \times 10 \times 10$

CELLS 10^{14}

DNA

EACH LIFE HAS 10^2 YEARS AT MOST

10 10 10 10

IV – *Ruler's Rules for Molecules*

How many atoms in a person? We are made of about 10^{14} cells. Each of them is made of about 10^{15} atoms. This means there are about 10^{29} atoms in any one of us.

We may reasonably have some difficulty in imagining such small sizes and such large quantities. Here, a royal personage is holding the yardstick by which all measurements are compared. He is the Ruler of that ruler, and so the yard is a measure of his own anatomy – from his nose to his outstretched hand.

His robes carry the subversive message of his unimportance, his smallness, relative to the very large number of atoms of which he is assembled. That is a fact that resonates with the irony of the atoms Lucretius predicted being very much divisible into their constituent parts.

The diameter of an atom's nucleus is only one part in 100,000, or 10^{-5}, of the diameter of its cloud. The volume of an atom is therefore the cube of that ratio, or approximately 10^{15}, or a million billion times greater than the volume of its nucleus. A single atom in turn, is only one part in 10^{15} of a single cell in the royal body.

What about cells and bodies?

A single human cell is only one part in 10^{15} of – no, not the royal personage alone – but a social group of about ten people; a jury, say, or a family. This coincidence of ratios serves to tell us that no person is truly alone for very long. We cannot emerge from infancy alone, nor can we expect to become fully human in isolation.

In recent years a second and somewhat disconcerting fact about ourselves and our bodies has emerged from research into the causes of certain diseases. Not even our bodies are alone: during the nine months of gestation, each person and his or her mother exchange cells across the placenta.

Enough of these cells survive for decades to assure that each of us is in fact a mosaic of cells from at least two different people. Every mother carries some living cells from each of her children, and every child carries some living cells from his or her mother.

V – *Anywhere Else but Here?*

Is there life elsewhere in the Universe? Would it be subject to these constraints? We do not know.

Another great and wholly unintuitive insight of Einstein, borne out by decades of observation: while the space-time continuum of the universe has expanded much faster than the speed of light, nothing within the space – time continuum itself can go faster than the speed of light. This caps our ability to know how big the universe may have gotten since the initial moment. We are sitting at the center of a sphere of light that has come to us at that limiting speed since the first instant, and any objects that might be situated further away than light could have traveled since that initial instant, are simply barred from our view.

Within this sphere of knowability, we can see approximately 10^{11} galaxies with approximately 10^{11} stars in each. So now, as we sit here on Earth, each of us made of ~ 10^{29} different atoms inside of a 10^{10} year-old universe of no fewer than 10^{22} suns, the big question this raises is, when is "now?"

That depends on where you are: there is no single "now" that we can speak of except in our imaginations. In reality, we are separated by the time it takes for light to get from one of us to the other. A Mars Rover's "now" is many minutes behind our own here on Earth, and when we sit around a table, each of us has a "now" that is a nanosecond or so ahead of the "now" of our tablemates.

The fixity of the speed of light means that we cannot know whether or not there is life now on any other planet of any star in the universe. What we might learn in the future, would only be whether or not life capable of signaling us lived in the past, at a particular time to the distance of the planet from which we received its signal.

I – *No Sun until Volume 20*

We sit today on a rocky, metallic planet composed of the atoms of dead stars, orbiting a middle-aged star born of the same stardust some 5 billion years ago. We share the gift of an imagination capable of creating and remembering creatures that that never were in nature, like Alice and the Cheshire Cat, first drawn more than a century ago by John Tenniel for Lewis Carroll's *Alice's Adventures in Wonderland*. Our sun is predicted to be able to shed its current light for another five billion years or so before turning into a red giant and then a white dwarf, then a supernova whose remants may well become parts of other stars and planets, billions of years from now. It is held by gravity in an arm of a galaxy born some ten billion years ago, in a universe born some 13.7 billion years ago.

Imagine the approximately 13.7 billion years since the beginning of space-time as being recorded in a set of books. If each book had 450 pages, and each page were to record a million years, then each volume would be the passage of 450 million years, and 30 volumes would come out to be 13.5 billion years. If we let the first volume have 650 pages, it will allow the whole library stack in the drawing to be a record of 13.7 billion years of time passing in that part of the universe where we are now.

The sun and the earth both formed by gravity from the stuff of exploded stars – the atoms of the elements – about 4.5 billion years ago, a time recorded on the pages of volume 20. The first evidence of life is found in fossils resembling bacteria, dated to dated to 3.7 billion years ago, in volume 22; it took less than a billion years for life to form and persist on Earth. The bindings on each volume are an index of the kinds of life that were here in the periods they represent. From volume 1 to volume 19 – that is, for times from the beginning of everything until a bit more than nine billion years later, or about 5 billion years ago – there is nothing on any of the bindings, which is to say there was no sign of anything at all around here for those billions of years.

II – *Volume 19 and the Evanescent Snowflake*

Life on Earth was a rather rapid development. Life's beginnings may not have left chemical or physical evidence, but only a billion years after the Earth cooled down from its formation to allow water to condense on its surface, the first forms of life to leave fossils did appear in those waters. What could have happened between the formation of the Earth and Sun in times recorded in Volume 20, and the appearance of the first bacterial life forms whose fossils we can identify, in later times recorded in volume 22?

Whatever it was, it had to have happened using the materials of nature – the atoms that Lucretius imagined – and those materials had to assemble in some way that would permit information to be stored in them, and for that information to be copied, indefinitely. Otherwise we do not really have any way to understand the history we find recorded in these fossils, that is, the survival of so many forms of life for very long times, much longer than the lifetimes of any individual.

Even in the earliest days of Earth, we can imagine survival of information through chemical structures that use the atoms and forces of the universe. We need only look at a snowflake. Each snowflake is famously unique, though each has an overall similarity of shape; they are almost all six-sided. Snowflakes begin as tiny crystals of ice that condense and freeze from the moistness in cold air. Depending on local temperature and humidity water freezes directly from water vapor into one of two hexagonal shapes, a flat six-sided plate, or a long, six-sided prism.

Each snowflake is different, because each has fallen from the sky through a unique pathway of slightly varying temperature and humidity. As a result each of its six sides grows out in a unique pattern of plates and prisms. And because the history of the flake is similar for all six sides, all six sides will take on the same unique pattern. So a snowflake wears its history in its shape, all of it inherited from its immediate past. It passes the first test for an inherited history in a chemical form.

What about survival of the information after the death of the individual snowflake? Is it alive? When a snowflake dies by melting, nothing is left of that history; it is lost forever. Friends from Alaska tell of the great sadness they feel as their glaciers melt and recede in this period of global warming. In that case it is not just a snowflake's structure that is lost, but also the gigantic aggregate mass of many snowflakes over many centuries.

III – *Volumes 21 to 26, Invisible Gardens*

Archeobacteria, the first known forms of life, appeared in the seas in the period recorded in volume 22. Their least-changed descendants still thrive in deep-ocean vents, where the heat and the gases emerging from the Earth's mantle provide the environment that most closely resembles the Earth's seas at the time of their first appearance. Bacteria living deep in the sea would continue to be the only known shape life took for another billion years, taking up the time recorded in volumes 23 and 24.

Then about two billion years ago the energy of the sun's light was captured for food by a sea-borne bacterium that used that energy to strip the atom carbon from the molecule carbon dioxide, building more of itself from the carbon while releasing oxygen as waste product. Photosynthesis, as it is called, changed the planet. Since that time, the oxygen levels of our atmosphere are at 20%, while the levels of oxygen on our neighboring planets are well under 1%.

With oxygen available as a source of energy, sunlight, which had been unnecessary for the archeobacteria, also became unnecessary for a new kind of ocean-living bacterium that was able to break down dead stuff and build more of themselves from it, using the energy of oxygen in a living, warm, stable version of fire, called aerobic metabolism. For these first two billion years of life aerobic, photosynthetic, and archeobacteria proliferated in the sea, their diversity being played out in their metabolisms, not in their shapes.

About two billion years ago, at a time recorded in volume 26, the benefits of cooperation gave a new, much larger kind of cell a chance to proliferate. An archeobacterium engulfed an aerobic bacterium within itself, and the two lived happily ever after. The aerobic bacterium and its descendants were so effective at generating useful energy for the archeobacterium that the latter grew to a thousand-times larger volume, and its closely-held aerobic bacteria became mitochondria within its cell body, a permanent source of energy for the new, bigger kind of cell, and its descendants.

A second engulfment by an archeobacterium of a photosynthetic bacterium led to the appearance of the green version of this sort of big cell, with mitochondria, but also with the green descendants of the photosynthetic bacterium, called chloroplasts. Volumes 25 and 26 record the billion years or so when life became a mixture of sea-going big and little cells, photosynthetic, aerobic and archeobacteria.

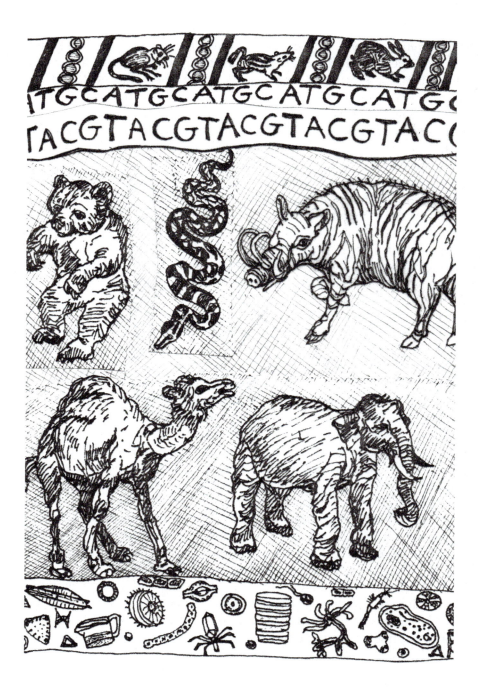

IV – *The Tapestry of Visible Life, Volumes 27 to 30*

Big cells soon gained the capacity for a more complicated, social form of existence. By the times recorded in volume 27, a billion and a half years ago, sponges and other simple multicellular forms of life appeared in the seas. And then, organisms made from many big cells appeared, whose ancestry was newly complicated by sex. Each had two parents, and each began as a single large cell formed by the fusion of a sperm and egg cell. This new way of making the next generation began in the sea, then found its way onto land, taking its place in, on and under the planet in a formidable Bayeux Tapestry of conquest by different life forms.

One sort of sexual creature, with a different head and bottom but with mirror-symmetric left and right sides, and with a symmetric backbone and four limbs, appeared some time in the period recorded in volume 29, less than a billion years ago. Other four-limbed creatures, tetrapods as we call them and as we might call ourselves after all, climbed out of the seas and occupied parts of the surface of the Earth not that long ago, well within the last billion years. And so opened the period recorded in the current volume, 30.

In the latest 450 million years recorded in the current volume, 30, big animals and plants, each following the sexual habit of having two parents, thrived for the period represented by the first 250 pages, until the great Permian-Triassic extinction, brought on by an undefined disaster of literally cosmic proportions, wiped out up to 96% of all marine species and 70% of terrestrial vertebrate species alive to that moment.

The survivors prevailed for another period represented by another 185 pages worth of time, until about 65 million years ago an asteroid struck the Yucatan peninsula. It so disrupted the Earth's land, water and air that all large life forms – including the dinosaurs, alas – were wiped out. Among the tetrapods that survived were small, flying dinosaurs whose closest relatives we know today as birds, and small, hairy creatures who raised their young on mothers' milk, the mammals.

How has the part of nature we call life managed to be so creative, so diverse, and so complicated over 4 billion years, able to overcome even asteroid impacts? How is it that despite periodic mass extinctions, life today is made up of so many different fertile species, each a separate population of individuals that – or in our case, who – can make babies with one another, and whose babies can make babies in turn?

V – *Imagination arrives only at the end of Volume 30*

The novelty of a snowflake cannot ever become the source of any form of historical memory longer-lived than itself. But persistent chemical memory of the past is the very hallmark of all of the forms of life that have filled the times recorded in the last ten volumes of our library in an unbroken persistence of complex life forms.

We humans appeared in Africa a few hundred thousand years ago. A note about our emergence would be on the last tenth of the last page of that last volume. And then, about sixty-thousand years ago, recorded somewhere toward the last sentence so far, would be the emergence in our species of language, texts and, in that mental world, thoughts of imagined and imaginary creatures like the Cheshire Cat and Alice of *Alice's Adventures in Wonderland*.

What have we and all these current and past forms of life got that a snowflake doesn't have? We and they all contain information, in the form of a text. Here, Charles Dodgson, who wrote *Alice's Adventures in Wonderland* under the pen-name Lewis Carroll, sits with his pet cat, wearing a T-shirt with Alice herself and the Cheshire Cat, his two most lasting imaginary creatures, both still alive to our imaginations, through his text.

The period punctuating the last sentence of Volume 30 would represent the past few hundred years, and so it would bring us to the present moment. The width of that simple dot records the time since science has taken hold of our imaginations as a way to understand all this.

In *On the Nature of Things*, Lucretius writes "Therefore, the supposition that, as there are many letters common to many words so there are many elements common to many things, is preferable to the view that anything can come into being without elementary particles." The notion of "many letters common to many words" takes us to DNA.

I – *Since Volume 21 , DNA Has Carried Information*

What sort of vine is this person climbing? Some time more than four billion years ago, in the period recorded in volume 21, a molecule called DNA came together as an assembly of atoms of the sort Lucretius was so prescient about. This molecule is still with us today because of the singular pair of properties it had and still has. It carries information in a sequence of 4 letters. In that, DNA is just like any other text. But unlike any text in any of our languages, these letters are assembled from only a few dozen atoms each. And unlike other text, DNA has the capacity to act at once as both book and scribe, making copies of itself.

DNA is assembled from only 5 of the 92 elements that occur naturally on Earth: carbon, phosphorus, hydrogen, oxygen and nitrogen. It is a molecule that always takes the same shape: a long skinny twisted pair of molecular vines. Each vine is a stack of copies of a single molecule assembled from phosphorus, hydrogen and oxygen (the Nucleic Acid or "NA" of "DNA," shorthand for "Deoxyribose Nucleic Acid"). Repeatedly jutting off each molecule making up the vine is a stem of the sugar Deoxyribose (the D of DNA) made of carbon, hydrogen and oxygen.

Attached to each sugar stem is one of four possible flat molecular leaves made of carbon, nitrogen, oxygen and hydrogen, called cytosine, guanine, thymine and adenine. As the intrepid climber has just noticed, the two vines of double-stranded DNA are held together by pairs of leaves.

Not all possible pairs of leafy bases will do. Only two pairs are seen. The same is true in DNA. The molecular versions of these leaves are abbreviated by the Latin letters C, G, T and A. Wherever there is a G on one strand, the other base must be a C, and wherever there is an A on one strand, the other base must be a T. The leaves reflect the molecular flatness of all four bases, which allows the two base-pairs to be flipped over so that they may be present as any of four base-pairs, G-C, C-G, A-T and T-A. These four may sprout from the two vines in any order, without disrupting their minimalist, smooth double-helical outer structure.

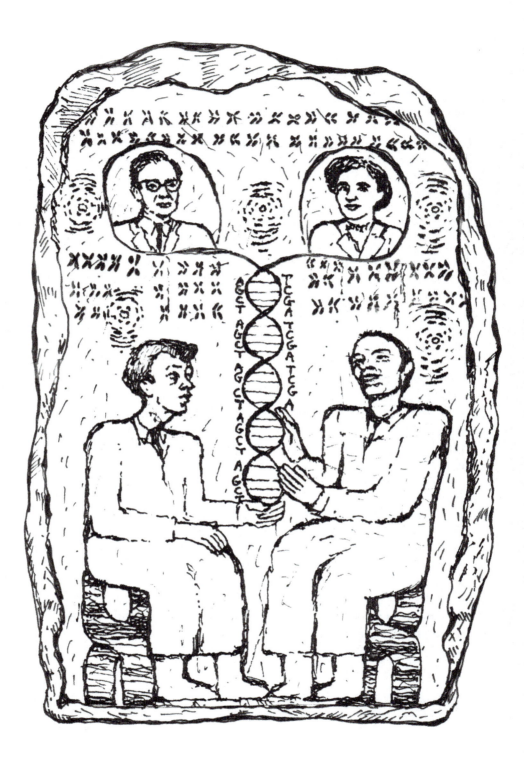

II – *In Volume 30, Watson, Crick, Wilkins and Franklin Explain How*

The discovery that DNA could have a structure that was self-copying and textual – each particular sequence of basepairs could be thought of as a sequence of letters – was reported in April 1953 as a 900-word prose poem disguised as a letter to Nature. The authors were the very young American postdoctoral fellow and fellow and naturalist James D. Watson, and the somewhat older British biophysicist Francis Crick, quite a pair themselves.

The paper, titled "Molecular Structure of Nucleic Acids," opens with a classic British understatement: "This structure has novel features which are of considerable biological interest." That's British for: "You'll be amazed; no insight like this has appeared in print since Lucretius two thousand years ago."

Watson and Crick wrote that because others had found clear X-ray crystallographic patterns from a preparation of DNA, the sugar-phosphate backbones separating the base-pairs have to run in opposite directions. Otherwise, a "crystal" of DNA would be a random assortment of molecules running one way and others running the other way, and no clear pattern would emerge.

They then drew an astonishing inference from earlier work of other, less biologically oriented, chemists: of all possible combinations of bases, only an A - T pair and a G - C pair had precisely the same distance to the two backbones.

Weaving these insights into the work of other scientists, they argued that the sequence of base-pairs was unrestricted by the structure of DNA, so that the two strands were both free to vary, while remaining informationally redundant. This, they argued, made DNA a candidate to be the chemical of inheritance, so that even the inheritance of forms through time might therefore be the property of an assembly of atoms.

Watson and Crick take their place as the formulators of this astonishing hypothesis. They sit as if they were in an Egyptian stele. Their bodies seem to resemble chromosomes, and the crisscrossed dots of the crystallographic pattern of a double helix are all about.

Hovering over Watson and Crick and embraced by the old strands about to be made into copies of the DNA below are Rosalind Franklin and her lab director, Maurice Wilkins. Franklin, who died of ovarian cancer in 1958 at age 37, obtained the crucial crystallographic photos of DNA that were the basis for Watson and Crick's structural models. Wilkins, who oversaw the crystallographic study of DNA, shared the 1962 Nobel prize with Watson and Crick.

III – The Radical Solution – Semi-Conservative Replication

Consider, Watson and Crick argued in *Nature*, what would be possible if the two strands separated, as the two vines do in this figure.

Each single vine would take with it a series of separated, single leaves. The sequence of kinds of leaves on either strand will be the same as it was in the double-stranded vine. Now consider what would happen if each single vine were to find itself in a blizzard of a single leaves, each stuck to a bit of vine. Any leaf from the blizzard might bump into any of the single leaves. But only the correct leaf for making a base-pair – AT, TA, GC or CG – would fit into one of the growing vines, adding to its double-stranded length.

In this way, each of the new double-stranded vines would reproduce precisely the sequence of base-pairs found in the original vine. Voila! From one copy of a text two are born, each from one parental strand of DNA. Each new double-stranded sequence would carry one of the parental strands and one new strand. Hence the boring political name given to the process: Semi-Conservative DNA Replication.

Semi-conservative replication resolves what had been a total mystery until its discovery. How can a chemical contain information and yet be supple enough to carry the same information through any number of generations of copying?

As Watson and Crick put it, "it has not escaped our notice that the specific pairing we had postulated immediately suggests a possible copying mechanism for the genetic material."

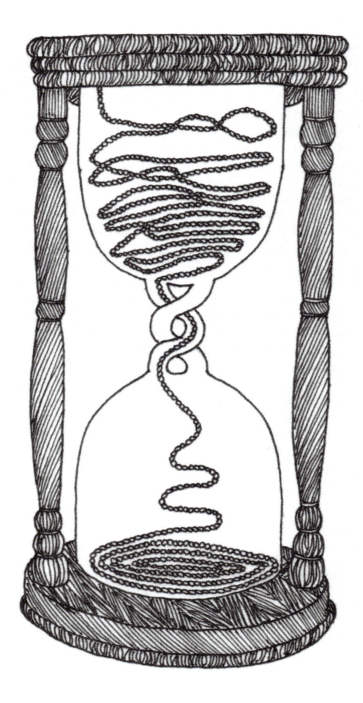

IV – Volumes 20 to 30, Information Stronger than Death

In this drawing a DNA molecule slips its way through the double-helical constriction of an hourglass, marking off the time since it was made from an ancestral strand and a set of new subunits.

Everything, Lucretius says, is just a particular, temporary local assembly of atoms that fill the universe. Nothing, he says, is anything but one or another such arrangement. On death, the arrangement that is a life simply ends, and the atoms of it go to form other structures. DNA is a source of astonishing stability in preserving information through time.

Nothing chemical prevents the existence today of a molecule of DNA, one of whose backbones may be billions of years old. All that would be required for a DNA with such an ancient backbone to slip through time indefinitely, would be that it happen to remain inside a living organism, and not be lost altogether to death.

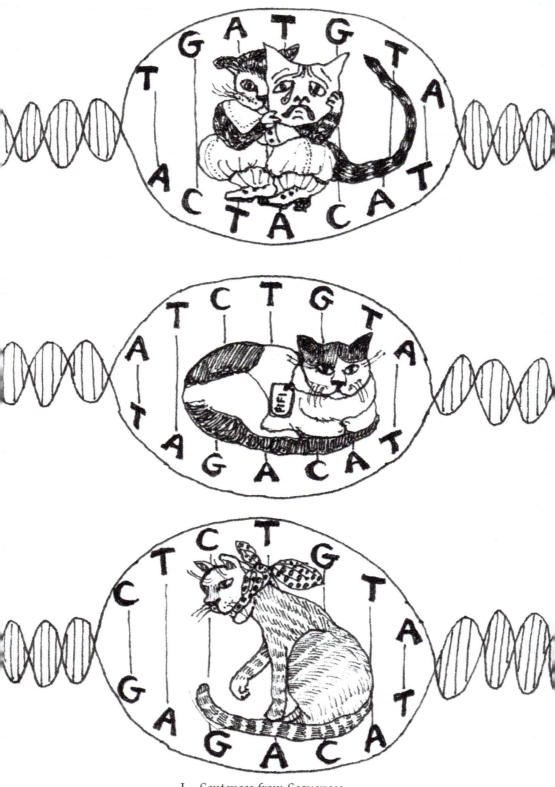

I – *Sentences from Sequences*

DNA is a text. Using a 4-letter alphabet of base-pairs it encodes information in its base-pair sequence in a manner identical to the way this sentence is encoding information using most of the 26 Latin letters and typographical symbols, like the commas used here, the spaces between words and the period at the end of itself.

Here, in each case one of the two DNA strands encodes a sentence in English, made of two sets of three bases, separated by a single base: ACT A CAT, TAG A CAT and GAG A CAT. Each is immediately translated – by us – into a meaningful sentence in English.

How is a real DNA text read by the cell it encodes? To answer we have to look in more detail at how DNA encodes the meaningfulness of its text into the construction of proteins. Proteins are strings of amino-acids; the amino acids that make a protein are a family of 20 molecules in which a single carbon atom links three elements: an amino group, NH2, made of one nitrogen and two hydrogen atoms; a carboxyl group, O=C-OH, made of one carbon, two oxygen and one hydrogen atoms; and a third group that determines the identity of any one amino-acid.

At a bare minimum, the DNA of a living organism must encode information for the construction of proteins to copy the DNA itself by separating the strands, then sewing in the completed second strands to make two copies of DNA from one. In addition, it must encode the totality of the proteins that assemble into the cells of the organism.

And also, somehow the DNA must encode the turning on and turning off of its protein assembly instructions, with variations both in time and in space. The regulatory information encoded by the DNA is more subtle, and it is not yet fully understood.

Our ability to study gene regulation is relatively new. For some time the classic gene was thought of as simply the stretch of DNA that encoded a protein. We now understand that all living things including humans are built from proteins encoded by DNA-sequences that are subject to regulation.

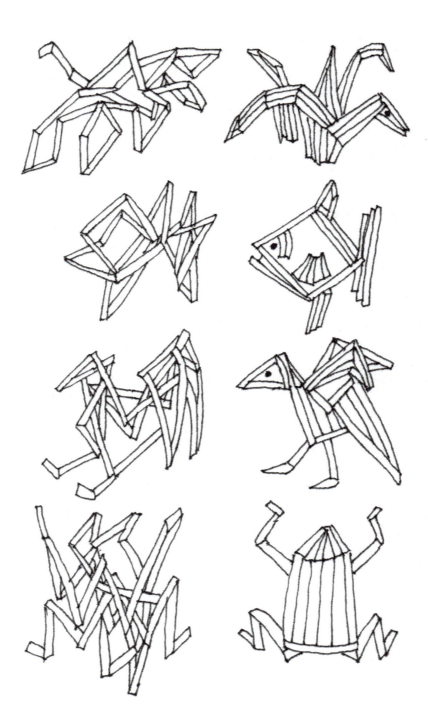

II – Proteins From Sentences

In the simplest and smallest of the 20 amino acids, glycine, the defining chemical group is just a hydrogen atom, but in other amino acids it can be complex, including several carbon and hydrogen atoms and perhaps also additional oxygen, nitrogen or sulfur.

Two or more amino acids can be linked in sequence, head-to-tail (amino group to carboxyl group), to form a string. Short strings are called peptide molecules. Proteins, such as hemoglobin, consist of long strings of hundreds, sometimes even thousands of amino acids.

An almost ubiquitous genetic code translates a sequence from the DNA alphabet of the 4 base-pairs into a unique corresponding sequence from this other alphabet of 20 amino acids. The code uses three-letter sequences of bases. Since there are 4^3 or 64 possible three-letter sequences, there is plenty of room to encode 20 distinct amino acids, and the code is in fact redundant. Some amino acids are encoded by two, a few by four, and one by six different triplets.

Even before the growing amino-acid chain is finished being translated from a stretch of triplets, it begins to fold up into a three-dimensional machine, like a molecular origami. The emergence of such a 3D, functional protein completes the translation of the meaning of a DNA sequence from the linear textual sequence of DNA base-pairs to the linear textual sequence of amino-acids.

While every protein will emerge from folding up as a single machine, any living thing big enough to be seen will be made of tens of thousands of different kinds of proteins, each interacting with many of the others. Different animals are often assembled from very similar proteins, but in most cases the sequence of amino-acids for a given protein is slightly different in two different kinds of animals, so each sequence may be thought of as characteristic of that kind of animal.

In this figure the folding of the molecular origami of an amino-acid chain attains the completion of the form of the organism that made it, whether bird or fish or frog.

III – *Not so Fast!*

Protein synthesis requires a middle step, translation, to carry information from the linear string of DNA base-pairs into the linear string of a protein's amino-acids. It begins with transcription, a variation on the theme of semi-conservative replication. A linear, single-stranded 1D, single-stranded nucleic acid much like DNA but not exactly so, called messenger RNA, is copied by base-pairing from one strand of a locally unwound double-stranded DNA.

Messenger RNA is very much like a single strand of DNA, but with a small chemical difference in the sugar backbone and with the base called T replaced by another base called U. When the messenger RNA is fully copied, it is released from the DNA which then re-seals as a double-helix.

In this figure transcription is shown in the Classical Greek mode with the messenger RNA embodied as the servant of an eternal Greek goddess. Here, he will take away the gift of meaning from a stretch of a long-lived DNA sequence.

A stretch of DNA encoding a messenger RNA is a necessary, but not a sufficient requirement for a gene. One form of gene regulation is messenger RNA censorship: sometimes an RNA transcript is cut to bits before anything further can be done with it. When it is allowed by gene regulation to be handed off, it is quickly bound to a big machine made of many proteins and RNAs called a ribosome. Firmly attached there, it is then decoded, as its specific sequence of bases is converted into a corresponding sequence of amino- acids.

Why this added middle step? One current idea is that the RNA step is a molecular fossil from an RNA world, a time before DNA itself emerged from RNA by precisely the sort of creative copying error that continues to occur in DNA today. We call these errors mutations.

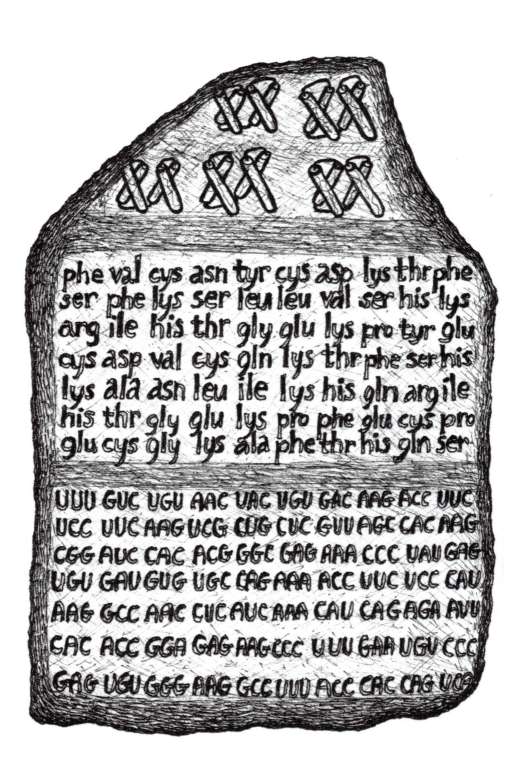

phe val cys asn tyr cys asp lys thr phe
ser phe lys ser leu leu val ser his lys
arg ile his thr gly glu lys pro tyr glu
cys asp val cys gln lys thr phe ser his
lys ala asn leu ile lys his gln arg ile
his thr gly glu lys pro phe glu cys pro
glu cys gly lys ala phe thr his gln ser

UUU GUC UGU AAC UAC UGU GAC AAG ACC UUC
UEC UUC AAG UCG CUG CUC GUU AGC CAC AAG
CGG AUC CAC ACG GGC GAG AAA CCC UAU GAG
UGU GAU GUG UGC CAG AAA ACC UUC UCC CAU
AAG GCC AAC CUC AUC AAA CAU CAG AGA AUU
CAC ACC GGA GAG AAG CCC UUU GAA UGU CCC
GAG UGU GGG AAG GCC UUU ACC CAC CAG UCG

IV – *Translating From the Language of DNA to the Language of Protein*

The Rosetta Stone was carved in Egypt about 2300 years ago when Egypt was a Greek colony under Ptolemy. The Stone, a remnant of a stele, was recovered by a French expedition a little more than two hundred years ago. It was taken by the British as spoils of war a few years later and is now residing in the British Museum.

The same text was carved into the Rosetta Stone three ways. At the bottom is a body of text in Greek, the language of Egypt at the time it was carved. The middle text uses the demotic Egyptian alphabet and vocabulary to spell out the same meaning. At the top of the Rosetta Stone ancient hieroglyphs were used to make the same text a third time.

Similarly we can read this drawing of a molecular Rosetta Stone from the bottom up, beginning with the DNA text of a protein-coding sequence having already been transcribed into a messenger RNA, here separated into the triplets of bases that encode different amino acids.

In the middle panel of the drawing, each triplet of RNA bases is translated into the particular amino acid it encodes. The names of the amino acids have been abbreviated to make the drawing legible; in full, they would be "phenylalanine valine cysteine ..."

The Genetic Code is the step from the bottom to the middle panel in this drawing. Both the bottom and middle panels show fragments of text. The text at the bottom contains only 35 of the 64 possible triplets of bases; but a much longer text would include all of them, each one many times. Likewise, the text fragment in the middle panel contains only 18 of the 20 amino acids.

At the top panel of both the original Rosetta Stone and this drawing the alphabetical texts below are translated into a language where symbols have entire meanings, much as in ancient Chinese. In the drawing the meanings of the sequences of bases are now translated into copies of the folded, functional shape of the entire encoded protein.

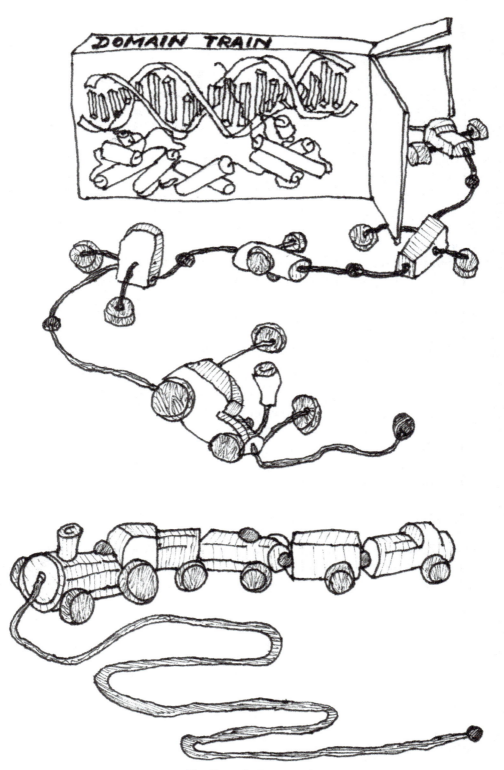

V – Domain Punctuation

Our own human genome has about 3 billion base-pairs. The protein-encoding parts make messages of various lengths, but at the high end a protein will not be much more than 10,000 amino-acids long, and a more typical size for a protein is in the range of 1000 amino-acids.

As a new string of amino-acids is forged on the ribosome, the final folding of the completed protein turns out to be a very elegant event. Different stretches of the nascent protein start folding on their own into one or another of a very small number of characteristic structures. These local origamis are called domains.

It is odd to find the same domains in so many different proteins, which raises the serious problem of likelihood. Given the enormous number of possible DNA sequences and therefore of amino-acid sequences why would sequences encoding domains be so common in proteins?

More surprising still, between the folded domains are often stretches of amino-acids that are flaccid, flexible, and undertake no specific folding at all. This allows the protein itself to assemble from its various domains into a single machine with a specific function.

The final function of the protein – a train in this figure – emerges first from the folding of the separate domains, then from the assembly of those domains into the finished protein.

How many of those typical proteins could our genomes encode? To get the answer, we divide the total number of base pairs in the genome (3×10^9) by the number of base pairs needed to encode a sequence of 1000 amino acids (3×10^3) to get 10^6. So if the only function of the base pairs was to encode proteins, as many as one million different proteins of 1000 amino acids could be encoded. But how many protein-coding sequences are actually in the human genome? The unexpected answer is about ten or twenty thousand, no more than that. This discovery that only a few tens of thousands of different proteins are needed to make a person means that the information for encoding proteins takes up no more than one or two percent of the genome.

I – *Circuits of Design Through Time*

There is a good reason why the sequences encoding proteins are so small a fraction of our DNA. All but one or two percent of our genome is given over to gene regulation. Most – but not all – of the non-coding DNA in our genome is used for regulating whether or not a protein will be made from any one of those ten to twenty thousand coding sequences.

Some stretches of the non-coding DNA are transcribed, not into messengers but to make non-coding regulatory RNAs that destroy specific messengers. Other stretches encode small RNAs that directly activate transcription of a messenger RNA; small RNAs can thus be either off- or on-switches.

In addition to the DNA encoding these regulatory RNAs, some of the proteins encoded by the genome also regulate transcription from a gene; sometimes it is their own transcription they regulate, sometimes the transcription of another gene, and sometimes the transcription of more than one gene. Finally some stretches of the genome serve in regulation simply by encoding a specific sequence to which regulatory proteins or RNAs may themselves bind.

The DNA of a person is organized in a sequence of texts much like the volumes of an encyclopedia. Scientists first saw the "volumes" of human DNA more than a century ago as separate human chromosomes. But because of gene regulation the DNA of a person is much more a set of programs than it is an encyclopedia, despite the coincidence of the 11th edition of the Encyclopedia Britannica coming in 23 volumes and a person's DNA being separated into 23 pairs of chromosomes.

In this drawing, the chromosomal location of a given stretch of coding DNA is no more than a bookmark for regulatory purposes. The DNA regulatory programs for different tissues are set aside in different volumes, but not at all the same as the chromosomes in which they are found. The circuitry of regulation can reach across the entire genome to turn sets of genes on or off, so in regulatory terms the DNA encyclopedia is totally interactive.

II – *Circuits for Many Tissues, In One Living Thing*

All animals and plants began as single, fertilized-egg cells. The egg cell's guidance of the construction of a body from a single fertilized egg cell is a deep, subtle library of regulation. In the final differentiated part of the body, different proteins will be synthesized in different cells. The initiator of all these cascades of differential gene expression in tissues, is the immediate differences in gene regulation that occur in the daughter cells after fertilization of the egg. These differences in gene expression emerge with the first cell divisions after fertilization, according to the luck of the different regulatory molecules each new cell would capture from the fertilized egg.

The initial regulatory program for making a multicellular organism is placed in the egg cell before fertilization, as that cell is itself matured by its own cascade of gene regulation. In the assembly of a fully differentiated egg cell, regulatory RNAs and regulatory proteins are placed inside the cell in richly asymmetric patterns. On fertilization and with the first cell division of the fertilized egg, these regulatory molecules are distributed differently in that fertilized egg cell's descendent cells.

There they open a variety of cascades of differentiation that need not be the same in the daughter cells. As these play out, lineages of cells – all of them carrying copies of the same DNA of the fertilized egg – become increasingly different from each other. Here the puzzle pieces of earliest embryonic differentiation into tissues are merged as they form one complete bird, with all its parts linked through later circuits of differentiation that maintain body structure and function over the lifetime of the bird.

Throughout life cascades of regulation also open to respond to many outside stimuli; we may experience these for instance as hunger or fullness; tiredness from lack of sleep or alertness upon awakening. Here hunger makes the bird alert and ready to pounce.

III – *Families of DNA with Great Similarity*

How do regulatory circuits arise? The novelty of a new DNA sequence can arise in many ways, not just by an error in the copying of a single base-pair. Sometimes a stretch of DNA is duplicated in tandem, then copied over twice, so that one offspring DNA has more than one copy of that stretch. That error occurs often, as an unintended consequence of the fact that the two strands of DNA run in opposite directions. It is an error that can build on itself, so in successive generations that two copies become four, four become eight, and so on. When the stretch of DNA so mishandled contains an entire gene encoding a protein, a large number of copies of that gene may accumulate.

In the large cells that appeared in the times recorded in volume 25, and then in the multicellular creatures assembled from large cells that followed soon after in the later times recorded in volumes 27 to 30, the genome within each cell had enlarged by gene duplication by about a thousand-fold, reaching billions of base-pairs per genome.

When mutation would then have its way, almost all of the copies in a gene family might be launched on different paths of mutational change over time while one copy could remain to provide the original function of the gene.

The baby is thus assembled from dots, because gene duplication and differential gene regulation together assure that every part of its body was produced by a circuit of gene regulation operating on families of genes to generate a pattern of proteins unique to that part.

IV – *Seeing Colors Emerge*

How do we see colors? Gene duplication has provided our genomes with a family of four proteins that together give us a colorful world in sunlight, and a black-and-white one at night. Three types of light-sensitive cells in the human retina provide the nerve signals for color perception. The three types differ in the protein portions of their respective light-sensitive molecules, called photopigments. Each shows a different peak sensitivity to the wavelengths of the visible spectrum of light. The human capacity to distinguish yellowish red from yellowish green, regardless of the relative intensities of the two, comes from the different peak sensitivities of two of these molecules, erythrolabe and chlorolabe. A third photopigment, cyanolabe, has a peak sensitivity in the blue region of the visible spectrum, and a fourth, rhodopsin, is active in very dim light, in which we do not see colors at all.

So the question then becomes, why does everyone have light-sensitive cells with only one pigment protein active in each, even though thanks to gene-duplication the family of genes for all four are present in every cell? As our brains and eyes develop, gene regulatory circuits assure that a cell in the retina that will express one of these genes will not usually express the others.

A radio is designed to be tuned to any one frequency so that one station may be heard without interference from the others. When gene regulation pushes one of the four buttons on this photoreceptor cell's genome, one light-sensitive protein's gene is chosen and the other three genes for photopigments are silenced throughout the lifetime of the cell. In the embryo different cells of the developing retina get tuned to different photopigments, so that each cell makes only one. The dial of gene regulation tunes this radio to the peak frequency of chlorolabe, so it will be best at picking up greenish light.

All four photopigment genes are members of a family that arose in the ancestral past by repeated duplication from a single gene. The most recent duplication made our red and green photoreceptor genes. That is why they are next to one another on the X-chromosome, while the other two are on other chromosomes. The blank buttons are waiting to be assigned new photopigments after further gene duplications.

LIZARD TORTOISE HUMAN

V – Similar Embryos, Different Outcomes

Any form of life big enough to see is made of many different cells, and fully differentiated into dozens or even hundreds of different tissues. But all the cells of a living person are the descendants of one fertilized egg cell and all inherit the same exact DNA sequence, always copied by semi-conservative replication. We have begun to understand how both can be true.

All specialization into different tissues occurs through cascades of differential gene expression that are laid down by these earliest cascades as the emergence of an ever more complicated embryo from that fertilized egg, as it eventually becomes a fully functional individual organism.

The DNA of a fertilized egg – whether of a lizard or a tortoise or a person – is a catalog of programs for different tissues, with each tissue assembled by differential expression, and with expression of some genes unique to each tissue. With about 10^{14} cells in each of us, most genes read out as proteins in only a small fraction of cells at any one time. For example, expression of DNA sequences encoding the protein keratin are regulated to be active only in tissues that face the outside world such as skin, nails and hair.

We have much to learn about the cascades of differentiation that allow a fertilized egg – the first cell of what will grow into a plant or animal – to regulate its DNA transcription so that its two daughter cells, and all their progeny, are launched into different cascades of differentiation.

Here three examples of such cascades have spilled out in three developing embryos, to yield a boy, the lizard on his shoulder, and the turtle in his hands. The wallpaper behind them shows the forms their respective embryos took as they differentiated from their respective fertilized eggs. Why did these three have such similar forms in their earliest embryonic stages, and could the answer lie in what the three still have in common, even as young adults?

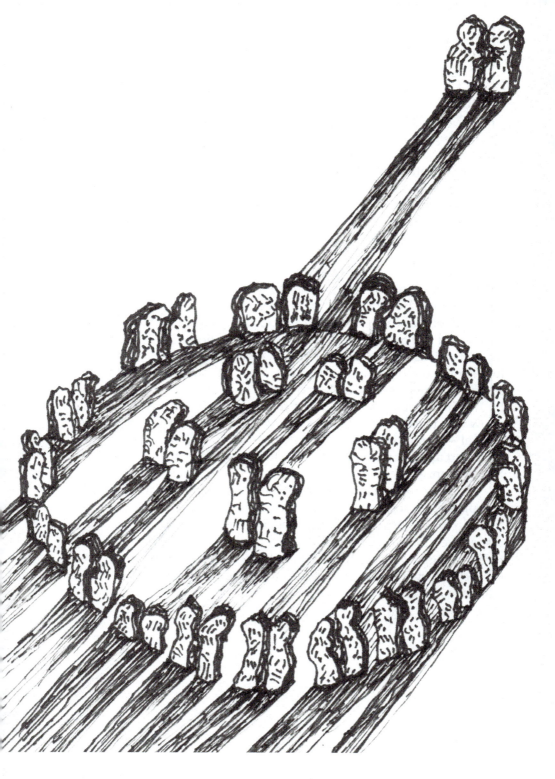

I – *Boy or Girl, Built From Both*

Big cells built from the information in billions of base-pairs of DNA package their DNA into segments – volumes – called chromosomes. Packaging is really too crude a word: chromosomes are tight coils of DNA wrapped around sets of proteins made from stretches of the DNA that encode them, with other regulatory proteins binding and unbinding the supercoiled state to allow or block transcription of a given gene.

Having two parents means there must be two sorts of cell division. The regular sort – called mitosis, simply grabs one of each copy of each chromosome and puts it into each of two new cells, each the genetic descendent of the original cell. Our bodies are built from mitosis coupled with gene regulation.

However meiosis, the sort of cell division that goes to make sperm, pollen and eggs, puts only one copy of each chromosome into those cells, in order that the fertilized egg not have more than two sets of chromosomes. In this figure, pairs of chromosomes have taken on their most condensed form, as their enormous lengths of DNA neatly coil up into visible, monumental forms. Here, they line up in the monumental forms great stones take on at Stonehenge, each chromosome in a pair from one of the organism's two parents.

Each chromosome of each pair is a single volume of the encyclopedic self-regulating text that is the genome of a given kind of multicellular life. The stones of Stonehenge were assembled there into a single instrument to capture the rays of the rising sun at the summer solstice. Similarly, no separate chromosome can contribute with the full expression of its DNA, except in the presence of all the others.

II – *Return of the Green Grandpea*

Well before DNA was imagined by anyone at all, the Austrian monk Gregor Mendel cross-bred two inbred lines of peas whose pods and fruit were of distinctive colors. He closely observed the offspring peas in their pods. One line bred true with yellow peas, the other with green ones. Mendel could mate two pea plants to each other at will, by dusting the flower of one plant with the pollen of another. On cross-ing – F1 signifies the first filial generation of sons, a gender-based term we are stuck with – the offspring peas were all yellow.

Mendel called the yellow color the dominant version of what we now call a single gene. The yellow color is the result of a single gene being able to make its protein successfully, while the green color re-sults from the failure of a single gene to express itself, either by muta-tion in the coding or in the regulatory portion of that gene's DNA.

A plant born from a pea that received one version of the relevant chromosome from a yellow-pea parent and one from the green-pea parent has yellow peas because it receives one copy of the functional gene, and one is enough to make the pea yellow. But this yellow off-spring pea is not like its yellow parent in one regard. It makes two kinds of pollen and two kinds of egg-cell: one kind contains the chro-mosome that has the functional version of this gene, and the other kind contains the other chromosome, which cannot express the yel-low-color in the plant.

When such "hybrid" yellow offspring plants are bred to each oth-er, each one of them has a 50% chance of donating a pollen or egg cell with a functional chromosome to the next generation, and 50% chance of donating one without a functional gene.

The result, as Mendel famously reported more than 150 years ago, is that on average one in four F2 "grandchildren" of the initial cross be-tween yellow-pea and green-pea parent plants, will receive two copies of the functional gene, two in four will get one copy of each version of the gene, and one in four will get two copies of the non-functional gene, and will be as green as its green grandpea was.

III – *The Creative Discarding of the Past*

What a waste of genetic information we trade for the novelties of sex at every generation! Egg, pollen and sperm must have one copy of any chromosome, and only one. That way the fertilized egg gets two copies for each chromosome, one from each parent. At each generation, having two parents means that the chromosome you got from your mother was one of two that she had gotten from her parents. Because you got that one, the other was lost.

And the same is true for the other parental chromosome of the pair: if you got, for instance, your father's mother's chromosome, you did not get, nor will you ever pass on, your father's father's version of that chromosome. There, in the instant of fertilization, half of the genetic information from each parent has been left behind.

What is the benefit of such a loss? You, and every individual that starts from egg and pollen or sperm, began life with a brand-new, never-before seen version of the DNA of its species. Not one, not two, but three shufflings of the chromosomal deck assure that happens at every generation.

First, before the sperm and egg are made, the chromosome pairs in the cells that will become sperm or egg line up and recombine, that is, they exchange DNA sequences. Second, every sperm, pollen and egg cell only gets one of the two parental versions of each chromosome; half the time that chromosome is itself a recombinant of sequences inherited by a parent from two grandparents. Third, each sperm or pollen and egg meets its partner entirely at random.

Recombination, random assortment, and random choice at fertilization together assure each member of a sexual species carries a truly novel version of its species' genome.

IV – *Blood, Not "Blood"*

Humans and peas both begin as fertilized eggs, so we are each products of a hand dealt by fertilization at every generation. That means that we each carry a random assortment of choices from the two parents, for every gene. Here's a Mendelian example from our own families.

The red blood cells that bring oxygen to all the cells of our bodies have proteins on their surfaces that are encoded by two genes, one making a protein type A, and the other making a protein type B. Our red blood cells do not need either protein, and cells without either A or B are called type O. Like yellow for peas, blood-types A and B are both dominant; like green for peas, type O is recessive.

Since our parents may each have any two of these three possible versions of the gene for blood-type in each of their two paired chromosomes, they can pass many different combinations on to us. These combinations will be expressed as the presence on our red blood cells of one, or both, or neither of the proteins they encode. We call these different sets of proteins blood-types. There are four: A, B, AB and O, depending entirely on which versions we inherit from both parents for these two genes.

Consider the case of a woman with type A blood who has one chromosome with the gene for A and the other chromosome with the gene for O. She has two children fathered by a man with type B blood who has inherited one chromosome with the gene for B, and the other chromosome the gene for O as well. By the shuffle of egg and sperm formation and random assortment, along with the role of the dice at fertilization, their daughter has inherited the A from her mother and the B from her father. The son has inherited the O from both parents.

The mother exhibits blood-type A, the father blood-type B. And so, the hand this family has been dealt does not include the expression of either parental blood- type by either child. So much for inheritance by "blood," or "dilution."

V– *Birds and Bees; Girls, Boys, and Peas*

Let's go back and look again at those peas that Mendel studied. Like us, each plant begins as a fertilized egg cell. Like us, each plant's body emerges by cascades of gene regulation. But unlike us, each plant's body makes both eggs, and pollen. For a plant rooted in the ground, no second kind of plant is available to bring pollen to its flowers. Rather, each plant makes both the flower and the pollen within it, and depending on the plant, the pollen is either allowed to fall onto the egg-containing tissues of the pollen at random, or with the help of wind, or by the action of birds and bees. But in every case, the plant is complete, making both the pollen and the egg cells that carry one chromosome each for all chromosome pairs of the plant.

We on the other hand, like birds and bees, arrive at birth in one of two forms. One form carries the ability to produce eggs, and carry fertilized eggs through to the birth of an infant child. The other form makes, carries around, and can deliver the sperm. Encoding two different body types is a big burden on the circuitry of gene regulation.

In the female case, there are 23 exact pairs of chromosomes, one from each parental egg and sperm cell. In the male case though, one chromosome of the female pair called XX is not an X, but rather a stumpy, diminished version of an X called a Y. The Y begins the differential gene expression necessary to convert the body type of the embryo into that of a boy rather than the default form, which the full 23 pairs construct as a girl.

This is why boys more likely to be colorblind than girls. The genes for the erythrolabe and chlorolabe pigments needed for normal color vision lie next to each other on an X chromosome. A girl will inherit one X from her mother and one X from her father. But a boy will inherit only one X, from his mother. In populations of European origin about 1 percent of X chromosomes lack the gene for erythrolabe and approximately another 1 percent lack the gene for chlorolabe. These failed genes occur throughout our species, though they are less frequent in populations of Asian and African origin.

Assuming the frequency is 1 percent, a boy has roughly a 2 percent chance of being red-green colorblind. For a girl to be colorblind she must have the same defective gene on both X chromosomes: 10^{-2} x $10^{-2} = 10^{-4}$, or 0.01 percent, for erythrolabe, and likewise for chlorolabe. So, red-green colorblindness is approximately 200 times less frequent in girls than in boys.

I – *A Very Short DNA Has Enough Versions for Us All*

DNAs found in nature are each very rare. That's because like any written message, any stretch of DNA represents a single possibility out of a great many. How rare is any single stretch of DNA? With four base-pairs possible at every position in the string, the number of possible DNA sequences of any size will be 4, multiplied by itself the same number of times as the length of the DNA string. So any single sequence of length N base-pairs, is one of 4^N possible sequences.

The number of possible sequences rises much faster than the number of atoms used to make the DNA itself. A DNA stretch 3 base-pairs long is one of 4^3 or 64 possible sequences. A stretch 6 base-pairs long is one of 4^6 or 4096 possible sequences though it has only twice the number of atoms of the shorter one.

A single version of DNA information is so improbable because so many alternative sequences are equally possible, just as the improbability of finding another person with your name is the consequence of the very large number of possible sequences of English letters that could be assembled in a string as long as a person's full name.

How rare is the DNA of a gene, a chromosome, a bacterium, a big cell? Let's start small.

Imagine that a comic with a dull wit wears a poster-board that describes his performance: act a gag, act a gag, act. He tells you that that is also his name. He then tells you that you cannot have that name, because it belongs to him. He tells you not to worry, there's a sequence just for you, and a different one for everyone, just not this one.

And he then explains – he is not a good comic, but is a well-informed one – that his name is one of 4^{17} possible DNA stretches, each a different sequence of the four letters A, C, T and G. You do the calculation and find that 4^{17} is about 1.7×10^{10}.

That means there are more than 17 billion possible DNA stretches of that length, more than twice as many as there are people on the planet today.

II – A Slightly Longer DNA Can Grow in One of Our Cells

DNA sequences seventeen base-pairs long could be assigned to every person alive today and most likely to every person who ever lived as well. But naming people is not the way DNA works. It encodes information for transcription, regulation and translation. How long must a DNA sequence be to encode the survival of itself ?

Viruses are the shortest sequences capable of copying themselves and also wrapping themselves in protein coats for survival. They can be very short because they are parasites of the cells in which they grow, using the cells' ribosomes and proteins for their own replication. They are formidably aggressive, entering a cell and rapidly converting it into a factory for the production of themselves and then spilling out as the cell dies, ready to infect new cells for another round of murderous replication.

The information in a viral genome can be lethal to a cell that is so much more complex that a few billion base-pairs of information are necessary to keep it alive. So, as a bomb may destroy a house, the genomes of the smallest viruses are our most daunting competitors in the business of surviving.

DNAs for the smallest viruses, ones that can destroy a big and complicated cell the way these have, may have no more than a few thousand base-pairs. Their little genomes, smaller by millions-fold than our own genomes, are already capable of wreaking the havoc in *Guernica* – Picasso's painting of the bombing of a city – growing copies of themselves in the cell and then breaking it open.

How rare is any one of the smallest viral genomes?

Any DNA sequence long enough to survive as a viral genome can be viewed as just a single possibility out of a vast number of sequences of that same length. For the genome of a small virus, there might be about 6000 base pairs, so that genome is one of 4^{6000} possibilities with that same length. This is about 10^{3600}: a 1, followed by 3600 zeros. This number of possibilities is vastly larger than the estimated number of subatomic particles in the known universe (only about 10^{80}).

III – *A Person's Genome is Absolutely Unique*

How rare is a human genome?

Every cell in a person except sperm or egg has two versions of the human genome of about 3 billion base pairs each, rather than a mere 6000 for the virus. To write out the 4 to the three-billionth power number of possible human genomes in ordinary notation would require about 1,800,000,000 digits. At 10 digits per second, simply listing such a number of possibilities would take over 5 years of nonstop printing.

The DNA sequence of any of us is clearly unique in a very special way, emerging as it did from all those alternatives of the same length. Just one copy of it, though no longer than a meter, is one of so many possible sequences, that its uniqueness is hard to articulate.

We cannot even begin to know the possibilities opened by other sequences that long nor even to list them; they are effectively infinite in number. That is the natural underpinning of what every honest person has always known from direct experience: every individual is equally rare and equally precious and should be treasured while they are with us.

The particular sequences that make up each person's version of the human genome will not come about again. With 10^{14} cells each enfolding DNA that is about a meter long, the many copies of the unique DNA in any person's genome would be 10^{11} kilometers long if laid end to end. That would extend from Earth to the outer planets.

Compared to the paltry 10^{25} molecules of water in the mug of ocean water, a few of which would be fished out in the second dip into the oceans, no amount of random dilution and mixing would ever fish out a sequence even slightly similar to any of our DNAs from the sea of all possible DNA sequences capable of encoding a human being.

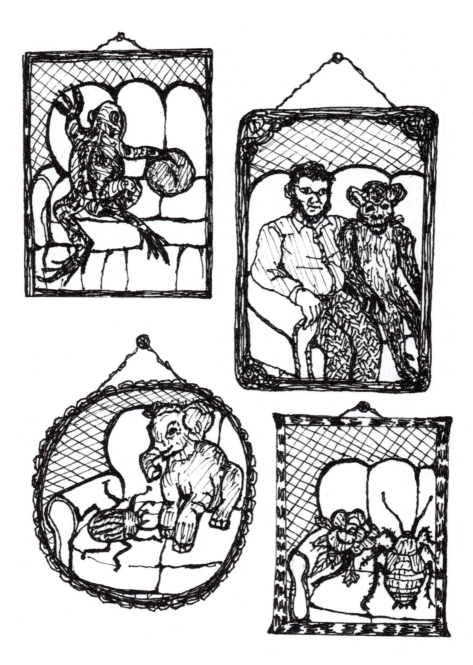

IV – *How Can DNAs Each Be Unique, Yet Be Similar?*

All of us are unimaginably rare, absolutely unique at birth. How then can it be that all living things share DNA sequences?

Given the incomprehensible rarity of every sequence big enough to form a genome, and given the existence of hundreds of millions of species today, each with its own species-genome and a unique version of the species genome in each of its members, we might presume that the genomes that characterize each of today's species would be completely different from one another.

About twenty years ago this presumption became testable by direct sequencing of the DNA of many species. Three discoveries stood out from the very first experiments.

First, all species' DNAs contain regions of similarity or identity with all other species' DNAs. That is a bit surprising, but could easily be explained by DNA traveling from one species to another, carried along in the genome of an infectious virus or bacterium.

But second, species that are as incommensurately different as frog and peach, beetle and elephant, roach and rose, or even chimp and professor, have overall sequence similarities in their genomes that range between 98% for the professor and the chimp, down to a small but real overlap of sequences even for the roach and the rose or the beetle and the elephant.

Third and most remarkable, the few percent of the DNA of each species given over to protein coding are much more similar from species to species than the sequences between coding regions that are involved in regulation. For instance, taking our own version of a given protein as the model it turns out that the amino-acid sequence of a protein is the same to within a tenth of a percent in a chimp's DNA; the same within 1 percent in our DNA and in the DNA of a dog; and 60% similar in our DNA and the DNA of a fly.

So we are presented with a real paradox: how can DNA sequences in all these species and by extension in all species be related to one rather than being distributed evenly though the vast informational richness of all possible long DNA sequences?

The solution is simply to beg the question and say: the DNAs of all species are related to one another. And then, because of the fact of semi-conservative replication of DNA, the penny can drop: if all current DNAs are related to one another, it becomes possible to imagine that all current species have a common ancestor.

V – Novelty, The Gift of Error

How might all genomes in nature be truly related; that is, be derived from common ancestral sequences? Let's follow what happens when DNA copying makes a mistake. For example, what would happen after two DNA strands separated, if one of the two new double-stranded copies put in a C opposite to an A, instead of putting a T there?

The C-A "pair" would be a counterfeit unable to assemble the two backbones properly. But that is not quite right: they would cause the new DNA to fail to maintain its absolute fidelity to the form of an undeviating double helix. Mistakes happen; how is it that DNA escapes these mistakes? Consider this remarkable possibility: that included in the information encoded in a stretch of base-pairs, one may find information that helps copy the DNA itself.

The Egyptian scribe in this figure is an enactment of that idea. Encoded by the DNA he copies, he personifies that use of the information in a DNA sequence. He is not passing on the text with absolute veracity; the DNA encoding him has not encoded perfection. He can detect his mistake and correct it, repairing the C-A typo. But there are two ways for him to make that correction: returning to the T-A base-pair, or "fixing" it by leaving the C and building from it a C-G base-pair.

Putting in a T for the C restores A-T base pair; the new DNA has a sequence identical with its ancestral backbone, and all is well. Putting in G for the A, however, does something radically different: it puts back a proper base-pair and the DNA is as sound as one could ask, but the sequence of bases has been changed!

Changes in base-pair sequence are called mutations. There's a common misperception that mutations are bad, or weak, or failures. In fact, they are merely textual novelties and only time tells what the consequence of the mutation will be to the DNA.

As a text, DNA must in some fashion have been able to encode information to accomplish its own copying, while maintaining the risk of novelty. Novelty – mutation – may be a typo that has no effect on the meaning of the DNA text, or one that ends the meaningfulness of the text, or one that adds a new meaning to the text.

I – *The Positive Feedback Loop: What Works, Persists*

Of all new meanings that may have emerged by mutation in any ancestral DNA, those that contributed to a new way for the DNA to enhance its own copying, would thereby have acquired a novel capacity to survive. As DNA copies itself by semiconservative replication, a change in sequence may be preserved and then copied into the future, until it is changed again by an other mutation.

Let's consider how this might work in terms of semi-conservative DNA replication. The copying of a DNA sequence requires a lot of energy and many machines made of protein and RNA. Genes for these machines are, of course, among the sequences encoded in any DNA capable of self-replication; that's how that capability is gained.

Now consider the next question: errors and their repair are making a host of new sequences over time, as the encoded machinery for copying DNA bumbles along. What if one or more of those mutational changes improves the quality of the copying?

Maybe it makes the copying go faster; maybe it makes it more accurate; maybe it just makes it work better in some other, undecipherable way. The long-term consequence of any novelty that enhances its own copying is to give the novel sequence a leg up into the future. Over time, all other things held constant, that new sequence should replace its ancestral one as well as all others that cannot compete for the copying of themselves. In this figure posting such a mutated DNA allows its novel sequence to copy itself in the mailbox with unexpected, novel efficiency.

This is the answer to the diversity of life in a DNA-based living world: a positive feedback loop rewards any sequence change capable of enhancing its own replication, and gives it hegemony over all ancestral sequences and other variants arising at the same time. It becomes the new normal sequence.

II – *With Time, Whatever Worked May Cease to Work*

How does the notion of this positive feedback loop apply to us and the other species of things big enough to see, the species that first arose from big cells and gained the additional sequence shuffling of sexual reproduction soon afterward? The ancient loop would be expected to reward DNA sequences involved in survival of a species genome by giving individuals born with them a preference at survival under some particular set of conditions. But consider the possibility that would arise whenever a change in those conditions meant that a small population within a successful species became physically or behaviorally separated from the rest.

The loop would continue to operate in the larger population, and also in the separated one. With time the changes that contributed to survival under these different conditions might be expected to diverge in the two populations. Consider next what would happen if the loop rewarded one set of changes in DNA with survival in one subpopulation, and another set of DNA changes with survival in the other.

Eventually there might come a time when offspring of parents who each belonged to a different subpopulation would be unable to survive and breed in turn, due to the differences in DNA sequence accumulating along different trajectories in their parents' sperm and egg. Each sub-population might still be the source of its own offspring, each would still be able to mate and bring forth a new generation in turn. That is to say, that a species might eventually be the source of a new species.

Of course, "normal" depends on where you live. So, sequences whose links to successful DNA survival were only distantly linked to its replication might also contribute to the survival of the life-form they encoded.

Any change that would allow a version of life to survive, would survive within that version of life. Ancestral DNA is a track, with a switch-point in the railway for the survival of two different life forms. The set of DNA changes an ancestral reptile would have to accumulate to run down the featherless-dinosaur track would preclude going down the feathered-bird track.

III – *The Asteroid and the Jackpot*

Charles Darwin imagined this consequence of a positive feedback loop that selected for different variations under different conditions more than 150 years ago, a hundred years before DNA's capacity to copy itself was understood.

His initial report of this insight, published as a brief book in 1859, gives the entire story away in its title: *On The Origin of Species by Means of Natural Selection, or the Preservation of Favoured Races in the Struggle for Life.* Not on the origin of life, about which he had no observations from nature, but the origin of species. By that long title he means to say that today's living species had an origin; they are not eternal. In that title he claimed for the living world a radical changeability.

The origin of a species had to have been an earlier species, one now dead and perhaps fossilized in the ground. His idea is presented here in a way that honors his prescience while recognizing the positive feedback loop in DNA replication that is its mechanism.

The feedback loop of DNA change that enhances DNA survival is represented here by a slot machine. The handle of the slot machine is a chromosome and the spinning wheels are ancestral species. They have rolled and come up with three ancestral mammals, from a species that might have died out a hundred million years ago.

The elephant, boar, anteater, dog, seal and bat are a sampling of the winners of that initial roll of the DNA slot machine 65 million years ago. At that time, and for a few tens of thousands of years afterward, the dinosaurs died from the impact of a giant asteroid that hit the Yucatan peninsula of Mexico and sent up a cloud of dust and ash that darkened the skies for a long time. Through that time and thereafter, mammals could survive and expand their range. We are members of the current set of survivors, today's species of mammals.

IV – "I Think …. "

We honor Darwin's profound insight by showing here a facsimile of Darwin's own notes from the moment of his understanding. Here, his note to mark the moment is not mounted as a page in a book, but as the face of a clock. That clock was ticking for more than 13 billion years, but its face was hidden until the moment of this drawing.

He writes "I think" and then draws a tree. The base of it, numbered "1," is a hypothetical ancestor of all the species that make up the tree.

The length of a limb is the length of time since a branching.

Each branching is the emergence of a new species from an older one.

All the species alive today have a crossbar; the other ancestral ones at every branching are dead and with luck remembered both as ancestors and as fossils.

To look at the species alive today, B, C and D would appear recognizably similar, but A would be quite different; think of a Boy, a Chimp, a Dog and an Albatross.

We see that B, C and D all have a recent common ancestor at the base of their branches. A on the other hand is not in that cluster, but it does share a common ancestor with the other three living species.

The generality is stunning: all forms of life are mortal. No matter how different, they share their current viability and their common ancestry but only the latter is permanent.

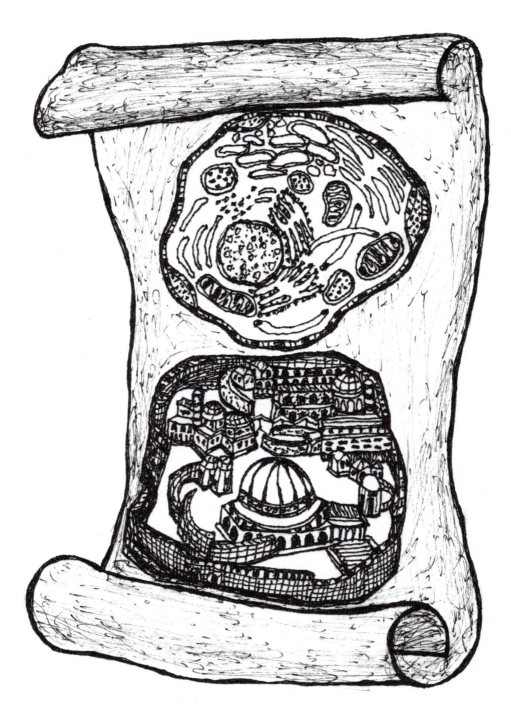

V – *Nothing That Works is Discarded Just for Being Old*

Each new species must have a genome very reminiscent of the one it had branched off from. Any single cell of any species has in its nucleus a genome that encodes not only the information for the construction and viability of an individual of that species, but also a record of its ancestry, in which those sequences that continue to work well enough under its own particular circumstances are preserved.

We see our species' history retained in the construction of our own bodies: early human embryos and early fish embryos look remarkably alike, because the parts of each genome given over to the regulation of gene expression in early embryogenesis have been preserved for a billion years or more, since we and fishes had one last common ancestral vertebrate in the seas.

No random mutations have survived from the time of the last ancestor shared by humans, lizards and turtles, that would have altered the early blueprint of embryonic development in each species, that sets up the vertebrate body-plan of bilateral symmetry with a head, a tail and four limbs. That is why the early embryos of a boy, a lizard, and a turtle, are so similar.

The generalization is simple: every nucleus of every cell of everything alive today is a repository of past wisdom in the form of texts that continue to inform the present. In that, a cell's nucleus is like the site of the ancient Temple in the walled city of Jerusalem.

I – *A New Leaf on the Long, Tall Tree of Life*

Darwin's vision of a single tree of life is DNA's positive feedback loop applied to all species. Darwin's tree of common ancestry has stretched and filled out by 150 years of study. Darwin had no clue that for almost all of the time of life on Earth, it had taken the form of bacteria, small cells, too small to see. The trunk of his tree labeled "1" turned out to be a long one, with branches representing life forms big enough to see crowning the bacterial trunk only in the past billion years.

Today we know the tree is rooted in the archeobacteria, whose ancestors are about four billion years old. Most of its foliage – the species alive today – are made of species still too small to see. In that sense, what worked first still works best.

Off one branch are all the animals big enough to see. A squirrel runs down the branch of mammals. The branch with no living species – leaves – held the dinosaurs; branching off from its base is the twig of today's birds.

When someone asks, "when does life begin" the answer is, "It never begins, it only ends; life having begun once has persisted until now." Lives however do end, and that is the price exacted by that kernel of regulation, the positive feedback loop that rewards the successful genome with survival.

Each leaf of a species emerged from an earlier species when mutational differences in its DNA led to enhanced survival of members of the species under some particular set of circumstances. As Darwin saw, those circumstances include the interaction of members of every species with all others in a given time and place; today we call the study of particular times and places and the species interactions occurring there the science of ecology.

This tree has grown, therefore, through a mindless process that rewards not goodness nor kindness, but only fecundity and survival itself. And where does that leave us? As a species, we're one of the estimated hundred million species alive today, an ordinary living twig on the great bushy top of the long spindly tree of life.

II – *Our DNA is a Palimpsest of Our Ancestry*

Differences in sequence accumulate in a genome over time. By comparing genomes of different species alive today, we can measure how different they are. If the positive feedback loop of natural selection working to preserve some mutations in DNA sequence through semi-conservative replication is a sufficient explanation for the Darwinian model of the origin of species, we can make a dramatic prediction. We should find that the more different two species' DNAs are today, the longer ago it was that they had the same common ancestral species.

For a stretch of DNA encoding a protein, not all novelties of mutational change in a single base-pair need necessarily have any effect at all on the final functional protein. The ones that have no effect because the genetic code is redundant are called neutral mutations. They will accumulate in any DNA over time without effect, and so provide a clock by which to measure how long ago the DNA molecules from two different related organisms might have diverged from a single species genome.

All ancestral species are dead, so with very rare exceptions their DNAs are not available to us. We can only compare the degree of similarity of the fossils of ancestral species with the current bodily structures of today's species. The comparison remarkably and wonderfully confirms the prediction. It is indeed the case that the more different two species' genomes are, the older the fossils of their last common ancestor species are. This allows us to draw the tree of common ancestry informed both by fossil and DNA evidence.

Nested first within the doll of our species, *Homo sapiens* – the thinking species – is another primate species' genome, to show that our two are most closely similar because we share a most recent common ancestor.

Nested inside that primate genome is the genome of a bird, a more distant cousin. Inside of all of them is the genome of a reptile and then a mammal, to represent the period of hundreds of millions of years when ancestral species of both lived together, with the reptilian species so much larger that we think of that period as the age of dinosaurs.

And then as the dolls get smaller and the tree fades into the past we find three nested genomes of an ancestral insect, fish, and arthropod. These last two are nested around a paramecium to represent the large single-celled forms of life from which all the others then emerged in not much more than a billion years.

III – *What of Our DNA is Just Human?*

Let's go back to that outermost, biggest Russian doll which represents our own human genomes. Measured against the 2 billion or more years of life the entire set of dolls represents, our genomes are remarkably new, and our numbers are remarkably large. The last ancestor common to all of today's primates – including us – died off about 35 million years ago. Our first primate ancestors did not appear until the last 1 or 2 percent of the time since life had emerged on Earth.

Our closest living primate relative is the chimp. Humans and chimps are built more or less from the same protein bricks and stones; it is the architectural plans of regulatory DNA that have been changed by natural selection since our last common ancestor species died off about 7 million years ago.

Seven million years is a very short time for those regulatory sequence changes to have been selected. The result of this burst of concerted selection has been to give us DNA sequences in regions that have remained unchanged since the last common ancestor of birds and mammals. Those sequences that are changed in us and otherwise the same in all other mammals and birds give us particularly human inherited qualities.

Such anatomically inherited qualities are easy to tick off: a bigger brain; a brain that matures slowly, leaving the human infant helpless and in need of close caring for years rather than weeks; a set of hands with thumbs that allow complicated manipulations of objects; a changed anatomy of throat and mouth that allows for a vast complexity of sounds we call speech; and most significantly, a new positive feedback loop: the mental world that emerges in each of us through experience, and that links them together.

In this drawing a pointillist baby sits surrounded by the chromosomal information from which it grew by gene regulation, happy with its toys and perhaps thinking its first unencoded thoughts.

IV – *The Sole Surviving Hominid*

Our enormous success at forming multiple overlapping social groups for the teaching and learning from each other throughout life has enabled us to use these human-specific mutational differences to construct mental worlds of novel complexity. Our most recent ancestral species all lived in Africa and so did our first human ancestors, for most of those millions of years. Our burst of growth as a species began about sixty thousand years ago when small populations of humans left Africa to migrate north to Europe, central Asia, east Asia and then Polynesia. Eventually, approximately fifteen thousand years ago, humans migrated over the Bering straits to the Americas, making final landfall in the tip of South America in the past ten thousand years.

And now, here we are. Seven billion of us cover the entire globe and are changing it forever. We speak thousands of languages, invent music, art, science, law, philosophy, economics, sports and games, and religions. Our brains record memories that can be shared and so last a lifetime and more; we imagine thousands of reasons for every event we experience and every feeling we have.

Tempting though it is to imagine that this history is also one of progress toward some sort of equilibrium, it appears that the first fully modern humans were the survivors of a history of natural selection in which each pre-human species seems to have survived the death of some if not all of its predecessors. Our species accomplished that most recently after leaving Africa to cover the planet, with the death of the Neanderthal hominid species in Europe some tens of thousands of years ago.

Novel mental complexity has always meant progress in the technical sense that gives us the power to show off, even to fly over the remains of our ancestral species, but it has not always meant kindness and generosity.

V – *Mental Worlds*

Darwin counted our unique mental capacities among the rewards of natural selection for us as a species. So on a foundation of neural circuitry we have built temples in our mind, often with a personification, like Athena, of a wisdom we hope we may attain. What a great success natural selection has found in us!

Or, perhaps not. Our mental worlds are no longer subject to natural selection. In fact, they never were; that's the reason for our spectacularly disproportionate success at proliferating. We have broken out of natural selection's slow sieve of mutation; we can simply have an idea, and choose to act on it. No need to wait for natural selection's positive feedback loop to select for a DNA variation that permits that new idea. Any brain can have any idea, any time.

So the joke of our imagination is on nature, but as a consequence it is on us as well. Our minds can encompass nature, but they can also go beyond nature, and imagine things that need not be possible in nature. When we do that, we may escape nature for a while, but in the end, these excrescences of the brain will be brought back into alignment with the facts of nature by the fact of species mortality. If there is one place for imagination to generate wisdom, that would be the place.

I – *Smart is Faster than DNA*

Consider what can happen when you can change a genome to order, without waiting for Natural Selection. The sequence encoding a green fluorescent protein named GFP can be inserted into the genome of a fertilized frog egg next to the regulatory region of a protein made only in a single tissue of the frog. The frog will be born as happy as any, and will not care that when it is illuminated with blue light that tissue fluoresces a bright green.

That DNA modification was done a while ago, before computers had flat screens and before those screens were also in everyone's pocket. A frog still cannot edit its own DNA, but now a person can. There are lots of reasons for the difference. But the one of most interest to us is our ability to manipulate texts and meanings in our minds. Our minds are much faster than our DNA.

Let's take this sentence: "A sequence of base pairs encodes the data of a gene storing information on the text of life," and mutate it, one letter at a time, to see if we can go from base to data to gene to text to life without ever losing the meaningfulness of the intermediate words. That would be a fair representation of natural selection's capacity to go from sequence to sequence, rewarding the new meaning by its survival, without ever going through a lethal mutation to meaninglessness.

When I taught this notion in a first-year class at Columbia University, I gave the following string of mutational changes one letter at a time, as my best effort, my fastest move from Base to Life by mental, not natural, selection: BASE bale dale date DATA date Dane cane cant cent gent GENE gent tent TEXT tent cent cant cane lane line LIFE. It took 21 mutational, meaningful steps, including a few back-mutations.

I told my class to send me an email with any faster path if anyone could come up with one. An hour later, back in my office, a student had sent me this: BASE bale dale date DATA date Dane dine dint dent gent GENE gent tent TEXT test lest list lift LIFE. This new path took 19 mutations, a ten percent shortening, and it took one hour to be thought through by a student who was not yet 20 years old. Within a few weeks a second, even shorter path arrived in the mail, sent by a student in a fifth-grade science class who had attended a similar lecture at his school.

There are 24 x 365 = 8760 hours in a year. The human generation time is about twenty years. So, human mental selection here was tens of thousands of times faster than natural selection.

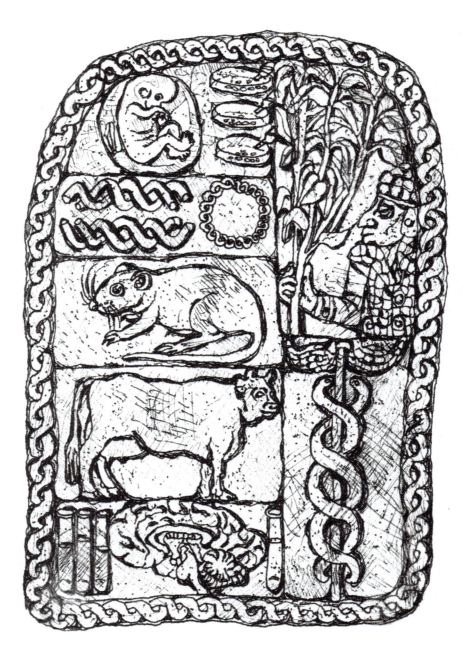

II – *Smart is Impatient to Change DNA*

The Mayan civilization began about ten thousand years ago with the arrival of the first humans in Central America. It was taken down by invaders from Europe about five hundred years ago. No doubt the invaders regarded themselves as more advanced than these Mayans who did not speak a civilized European language, nor worship a European notion of divinity.

But the Mayans were a religious people, and the drawing of a Mayan stele compels us to remember them and to imagine what they might have made of our current capacity to manipulate the sequences within the DNA of many species, including our own.

Ringed by the DNA double helix, the monument shows, on the right, the corn that humans had selected by their own weeding and breeding from a native grass. Below it, the emergence of a local medicine, informed by the human capacity to teach and learn about the risks and benefits of one or another plant or animal. And then on the left, a series of panels were carved to show the consequences of the human ability to manipulate the sequence of a DNA.

First, we see a panel showing the new human capacity to bring the fertilization of sperm and egg out of the body and into a dish. Just below it there is a representation of our forty-year-old ability to move pieces of DNA encoding genes of interest from one genome to another.

Below we see what appear to be a rodent and a cow, to represent a mouse and a cow that have received through DNA recombination in the laboratory the gene of a person or a fly or a bacterium. And finally, at the bottom, a representation of our ability – available in principle but unusable so long as we remember it to be wrong – to change the DNA of a fertilized egg in order to generate a baby with a novel mental capacity in its brain.

III – *Smart Can Play a Beautiful Tune*

For the past 11,000 years we have lived on a planet sufficiently stable that geologists call the period by a single name, the Holocene epoch. At the beginning of the Holocene humans numbered in the millions. We along with our domesticated animals made up about 0.1% of the total biomass of mammals. Today our numbers are about 10^3-fold higher at 7 billion, and approximately 80% of the mammalian biomass is made up of us and our domesticated species.

The technology of transmission of a string of sounds has changed very rapidly in the past century. The record player encoded a symphony or speech or song as a spiral track on a flat plastic disc. Like a stretch of DNA, that spiral could be decoded, but as sound through translation by vibrating needle into a set of electrical impulses which drove a vibrating cone in a speaker.

The recording of life has played out through the amplifier of natural selection for more than four billion years. We now have a lot of reasons to be concerned about its future. The Holocene Record Player has been superseded. Our species has for the past few hundred years shown an astonishing capacity to outstrip natural selection to create a version of Nature more suitable to our own species' immediate needs, at great cost to other species.

IV – *Smart is Not Enough*

Geologists have decided that our species' impact on our planet has been sufficiently degrading and irreversible to warrant the establishment of a new geological epoch, the Anthropocene. Thus begins Volume 31.

Markers of human-initiated change run through the Earth's water, air, land and life. In the past century the oceans have become more acidic as they take up the extra carbon dioxide we have put into the air by our appetite for the burning of fossil fuels. Global average temperature has risen in step with that additional CO_2.

All seven continents now show signs of erosion and intensive usage by our species, as woodlands, grasslands, steppes, shrublands, deserts and tundras become subject to our wish to manage, maintain and change them for our purposes.

These numbers mark both a geological epoch and also a change in tune. Here, Volume 31 floats on the rising seas of the Anthropocene, sharing ice floes with four threatened species: the Island Fox, the Polar Bear, the Lynx, and the Sea Otter.

Is the Anthropocene going to last 11,000 years as well? In geological terms it is hard to see how our planet can continue for much longer to carry the burden of our species' success at bypassing natural selection, whether expressed through our mental activity or our capacity for joint action at the largest scales, or both.

The adjustment, if it comes through natural selection, will undoubtedly be a problem for our species. After all, because Darwin's vision of natural selection still applies to us all even with our mental worlds, we will either survive as a species or not. Either way, chances are that DNA-based life will go on.

Our difference is, of course, that our mental capacities and our capacities for joint action do give us a unique extra capacity to control our future, a capacity we do not share with any other known creature on Earth. We are not merely a song on this record; we can make our own recordings.

Let's hope we use that power wisely.

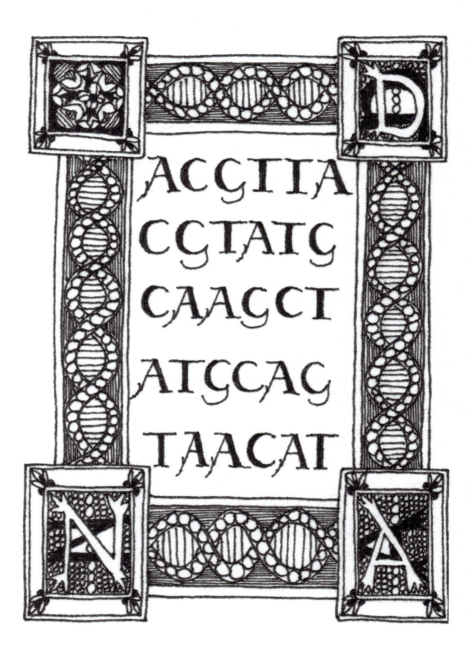

ACGTA
CGTATG
CAAGCT
ATGCAG
TAACAT

V – *Keeping the Music: the Inherited Capacity for Love*

The philosophers speak of four kinds of love, each having its place in the life of a person. Eros, for desire; Agape, for unconditional love; Filia, for family and friendship; and Caritas, for love and kindness to the stranger.

Let's close this book the way we opened it, with a thought about questions, and with the certainty that as no species will live forever, no question will have a final answer. Neither a DNA sequence in nature, nor any sequence we may construct and test in our laboratories or our farms will ever be fully knowable in advance of natural selection's capacity to favor it or kill it.

Does this mean we must give in to cynicism, or to despair? If there is one lesson we can take from our DNA sequences and the tapestry of our origins by natural selection, it is this: cynicism and despair are choices, but not ones encoded by our DNA. But these four kinds of love are encoded, and they can be expressed by any of us through our lifetimes.

None of us could have begun without Eros. None of us could have survived the burden of our inherited late-maturing neural circuitry without Agape from someone who cared for us at a time when we were of no use nor value except for our lovability. None of us could begin to understand the world and our chance of happiness in it without Filia. None of us can be fully human without Caritas, both from us and to us.

From that baseline gift of nature, it seems to us a small step to read the book of our origins in nature as allowing us – if not encouraging us – to choose to treat others only in the way we'd wish ourselves to be treated in turn.

We could not have drawn and written this book without the help of a number of people.

We thank Columbia Professors Darcy Kelley, David Helfand and Don Hood and course administrator Elina Yuffa for inviting us to assemble and present the very first version of this book as a set of three lectures to the first-year students of Columbia College in the Core course "Frontiers of Science."

We thank Catherine Macaulay, science teacher at Hunter College Elementary School, for asking us to present a shortened but no less rigorous version of those lectures to her class in what was an astonishingly intense seminar with two dozen ten-year-olds.

We thank our daughter Marya for reading the manuscript in its earliest version, her entire family for encouraging us both to see this through, and, in particular, our grandson for his editorial contributions.

We thank our student colleague Emma Chaves for the insight described in the legend to Figure 37 and for many incidences of kindness since.

We thank Summer and Connery Hart, Cynthia Peabody, Barak Wouk, Andrew Sinanoglou, Nathan Ashe, Miranda Hawkins, Susan and Ken Schept, Dawn Digrius, John Horgan, John McCaskey, John Loike, Pilar Jennings, David Kagan, Trevor Marshall, Gessy Alvarez, Alice and Rob Newton, Meredith and Michael Bergmann, Allison Levin, Katie Gerbner, and Sean Blanchet for encouragement and for reviewing earlier versions of the drawings and helping us to make ourselves clearer.

We thank CAL Dean and Stevens Professor of Philosophy Lisa Dolling and Columbia Professor of Psychology David Krantz who together also gave us good advice and in addition envisioned this book and its purposes even before we did.

We thank University Seminars Archivist Summer Hart for preparing both figures and text for publication and Stevens Professor Ed Foster for advice in publishing this edition, with support to both from Professor Krantz's remarkable gift to CAL.

We reserve for ourselves responsibility for all that is in this book, and hope the reader will join the circle of those who have helped us by letting us know of errors and misunderstandings.

Amy and Robert Pollack

AMY POLLACK graduated from the New York City High School of Music and Art, then earned an B.F.A. from The Cooper Union, and a B.A. from Brandeis University with a major in Art History and with honors in Graphics. Her recent works include illustrations for publications of the Stevens Institute of Technology's College of Arts and Letters, illustrations for the Columbia University student magazine "Sanctum," and illustrations in "Crosscurrents" magazine. In addition to illustrating articles, she enjoys making soft sculptures, tee-shirt designs, collages, and quilts. Her facility with scientific matters stems from summers spent at the Marine Biology Lab at Woods Hole Massachusetts, and on her work as a technician at Brandeis University, where she learned to make ribosomes from yeast. *The Course of Nature* is her first full-length book. An earlier version of *The Course of Nature* has been used as a reading in the Columbia College core course "Frontiers of Science," and in the Stevens Institute of Technology program in Arts and Letters.

ROBERT POLLACK graduated from Columbia University with a B.A. in physics, and received a Ph.D. in biology from Brandeis University. He has been a professor of biological sciences at Columbia since 1978, and was Dean of Columbia College from 1982–89. He received the Alexander Hamilton Medal from Columbia University, and has held a Guggenheim Fellowship. He is the author of *Signs of Life: The Languages and Meanings of DNA* (Houghton Mifflin/Viking Penguin, 1994), *The Missing Moment: How the Unconscious Shapes Modern Science* (Houghton Mifflin, 1999); and *The Faith of Biology and the Biology of Faith: Meaning, Order and Free Will in Modern Medical Science* (Columbia University Press, 2000). *Signs of Life* received the Lionel Trilling Award and has been translated into six languages. In 2010 he was elected to be the fourth Director of The University Seminars at Columbia.